# So You Want to Learn About Spiders

Written and Photographs by
Katrina Willoughby

I hear you want to learn about spiders!

All regions of the world except the cold Arctic and Antarctica are home to spiders. They are all around us!

Scientist think you are always within 5 feet of a spider and most homes have about 30 spiders living in them. Did you know that you share your home with so many spiders? Don't worry, most spiders run away from you if they can. They only bite people when they think they are really in danger.

Spiders are not insects; they are arachnids. Just like other arachnids, spiders have 8 legs. Scorpions, mites and ticks are other kinds of arachnids.

They all have 2 body segments. The segments are called the abdomen and cephalothorax.

Furrow orb weavers like this one can often be found on porches and other outdoor structures built by people.

Hairs on spider's feet allow them to cling to surfaces so they can walk on walls and upside down.

They don't hear, but hairs on their body pick up vibrations of sound waves. They may be able to tell when you are talking nearby.

There are two types of spiders; those that spin webs to catch food and those that hunt.

Hunting spiders track down and capture their food. They don't build webs. They have to have good vision.

Web spinners may have poor vision. This picture is of a spiny orb weaver spider. It can be identified by the spines sticking out from its body.

Eyes are in different places on different species. This can be used to help identify the species of a spider.  The mabel orchard orb weaver below lets dinner get caught in a web. It doesn't have to chase down its dinner.

Some spiders that don't hunt have fewer and smaller eyes than hunting spiders.  This web spinner has little eyes. A spider's eyes are designed more like human eyes than insect eyes.  Each eye sees a single picture.

Hunting spiders don't build webs. They live on the ground or in plants and have to chase down their food. One type of hunting spider is called the jumping spider. Can you guess why they are called this? They must catch their food by jumping on it and injecting venom.

The spider above is a jumping spider that has caught an ant. Jumping spiders may have the best vision out of all the types of spiders. Eight eyes are used by jumping spiders to look around and hunt for prey.

Hunting spiders have good vision and big eyes.

Jumping spiders leap to get away from danger. To jump, these spiders use special muscles in their upper body to quickly push a lot of blood into their legs.  The sudden rush of blood makes their legs go straight and sends the spider high. This is a unique way to jump. When you and I jump, we use muscles in our legs to move.

Jumping spiders can jump 50 times their body length. This spider is getting ready to jump away.

This green hunting spider is a crab spider. These spiders get this name because of their big front legs.

Below is a twin-flagged jumping spider. The name comes from the flag like appendages near its face. This little guy can live up to two years.

The wolf spider doesn't jump, it runs after its dinner. One is shown below.

Most spiders are nocturnal which means they are awake at night. Wolf spiders are hunting spiders that hunt at night. Sometimes you can see the reflection from their eyes as a flashlight passes over them. They chase down and grab their pray. They got their name because they hunt like a wolf. Wolf spiders can be found on the ground where they blend in with decaying plant material.

There are around 2300 species that are called wolf spiders. The arrangement of their eyes can help identify wolf spiders.

The female wolf spider will carry her egg sack with her. After the babies hatch, they climb on her back and she takes care of them until they are ready to be on their own.

Spiders hatch from eggs. The female will wrap the eggs in silk. The sack of eggs is placed in a safe spot.

For some spiders this means they place them in their webs. Many others will find a place to hang the sack.  This spider had her egg sack safely in her web. Like many other spiders, she will protect the eggs until they hatch. Some spiders even stay with their young after they hatch.

While hunting spiders don't make webs, some still spin silk. The silk can be used to wrap eggs, build shelters, or as an anchor when they jump or come down from a high spot.

Spider silk is stronger than steel or the stuff bullet proof vests are made of!

When inside the spider's body, silk is a liquid, when a spider expels the silk it turns into a solid.

There are different types of web spinning spiders.

The spiders that make big circular webs are called orb weavers. You don't find these in your home very often. Across the world there are over 3100 species of orb spiders.

Fossils suggest that orb weavers have been around for over 140 million years.

There is a spider called the Darwin that can have a web up to 80 feet across. Can you imagine a web that big?

Have you ever walked into a web? It probably belonged to an orb weaver.

Like other web spinners, orb weavers rebuild their webs every night. It takes about 60 minutes for a spider to spin a web. The old web is either eaten or rolled up and thrown to the side.

Funnel weavers make smaller webs with a funnel in them.  The spider often hides in the funnel. These webs can be found in the corner of your home but normally are outside. There are 116 funnel weaver species in North America and 1200 across the world.

They tend to get scared by sudden movement or light changes. They will run away to the back of their funnel to stay protected.

Can you see the spider in the back of this funnel web?

Individual spiders can make different types of silk. Some may be stronger or more stretchy depending on what they are using the silk for.  Some, but not all of the silk is sticky.  The spider applies the sticky substance onto the silk. The sticky substance is made in a gland on the spider's abdomen.

Spiders usually don't put this glue in the center of their webs. Tiny claws on the spider's legs let them travel along their web without getting stuck.

Weaver spiders tend to have longer legs than hunting spiders. This helps them from getting tangled in their own web. Daddy long legs webs aren't sticky. When prey gets tangled in the web, a Daddy long legs will rush over and wrap the bug with silk so it can't get away.

Daddy long legs are also known as cellar spiders. Their legs can be up to 2 inches long.

Almost all spiders are carnivores meaning they eat only other animals. A few species will enjoy nectar or pollen from flowers on rare occasions.  Out of more than 40,000 identified species of spiders only one eats a plant based diet.  That spider is a jumping spider in Central America and Mexico.

A few spiders are poisonous to people.

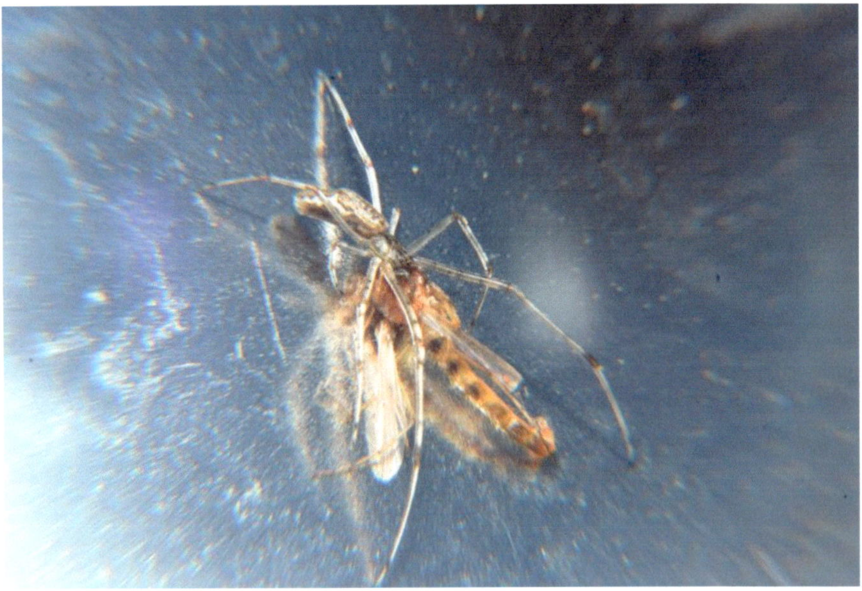

The first set of legs on a spider is used to taste its prey.

Spiders spit up or inject digestive juices into their food. These juices break down the food into a broth that can be drunk.

Here you can see a fly that has been wrapped up by this spider.  Have you ever seen a spider rush over to wrap up a bug that was caught it its web?  They can do this very quickly.

Arachnids don't have any bones. The exoskeleton on the outside of their bodies provides support and protection.

The old exoskeleton that the tarantula recently shed can be seen on the left in the picture.

So….

Did you learn new things about spiders?

I hope you learned new things and enjoyed this book!

This is the fifth book in the series "So You Want to Learn About..." Please check out previous books including the first book "So You Want to Learn About Butterflies" and check back for new books.

Thank you for reading.

## Other books in this series

So You Want to Learn About Butterflies

So You Want to Learn About Turtles & Tortoises

So You Want to Learn About Reptiles & Amphibians

So You Want to Learn About Insects & Bugs

Made in the USA
Las Vegas, NV
07 November 2026

57669931R00017